GEOLOGIC HISTORY
OF
ROWE, MASSACHUSETTS

by Norman L. Hatch, Jr.

Natural History Series
Number One

Rowe Historical Society, Inc.
1985

Library of Congress Catalog Card Number: 85-61223
ISBN 0-9614894-0-5

EARLIEST HISTORY OF ROWE

Although many would consider that the history of Rowe began a few hundred years ago when the first settlers arrived, or possibly slightly earlier with the activities of the local Indian tribes, it actually began around 600 million years ago with some very dramatic events that had a profound effect on the past few hundred years of Rowe history.

The topography of present-day Rowe is rugged. The generally hilly terrain is sharply dissected by Pelham Brook and its tributaries through the middle of town, by Steele Brook in the southwest, by Davis Mine Brook along the east edge of town, and by the spectacular gorge of the Deerfield River, which forms the town's northwest and southwest boundaries. Elevations above sea level range from 750 feet along the Deerfield River in the southwest corner to just over 2100 feet at the top of Adams Mountain. The topographic relief of 800 vertical feet in a horizontal distance of only 1000 feet between the Deerfield and Bear Swamp Upper Reservoir is certainly one of the more spectacular slopes in southern New England.

The town of Rowe lies roughly in the middle of the Appalachian Mountain chain, which is a geologic feature that extends at least from Alabama to Newfoundland. The present-day rugged topography of the Appalachians, including the hills and mountains of Rowe, results from the physical properties of the rocks, such as their hardness, plus the distribution of faults and cracks in them. These physical properties, in turn, are the direct consequence

Rowe Historical Society

Steele Brook Gorge—southwest Rowe.

of the major geologic events that have affected the rocks. Although geologists have long recognized that the rocks of the Appalachians have been folded, faulted, heated, squeezed, and injected with molten granite, it is only since the late 1960s that we have begun to understand the mechanisms and processes responsible. These mechanisms have come to be known by the general term Plate Tectonics.

Think of the earth as a ball of rock. The outermost few tens of miles of this ball constitute a layer, analogous to the skin of an orange, which is called the crust. This crust consists predominantly of light-colored rocks (for example, granite, gneiss, schist, and sandstone) where it forms continents, and of dark-colored volcanic rocks (mostly basalt) where it underlies oceans. We now recognize that, unlike the orange skin, the earth's crust consists of many discrete pieces, or segments, that vary in size from a major continent or ocean basin to pieces as small as a few hundred square miles. These pieces, which geologists call plates, fit together like pieces of a spherical jig-saw puzzle to form a complete skin. Driven by forces from deep within the earth, these plates are constantly moving relative to each other at rates of a fraction of an inch to a few inches per hundred years. In some cases, adjoining plates slide laterally past each other; for example, the Pacific Ocean plate and the North American plate are slowly sliding past each other along the San Andreas fault in California. In other cases, adjoining plates move directly, or diagonally, toward each other, which is what is currently happening between the Pacific Ocean plate and the South American plate. In such cases, one of the colliding plates may be forced down to considerable depths beneath the other. When this happens, the rocks of both plates become crumpled, folded, and broken (faulted), and the rocks of the underriding plate become intensely heated and squeezed. Spectacular mountain ranges and intense volcanic activity commonly result from such processes, and the modern Andes along the west coast of South America are an example.

The character of the rocks and rugged terrain of the present-day town of Rowe, as well as the pyrite, copper, talc and soapstone deposits, all resulted directly or indirectly from plate tectonic processes.

The other fundamental geologic processes involved in Rowe's earliest history are erosion and sedimentation. Many plate tectonic events, particularly collisions between plates, result in the elevation of areas of the earth's crust to form plateaus, ridges, and mountains. The actions of wind, rain, and freezing and thawing on these elevated areas tend to break up or slowly dissolve the rock. Numerous holes in a ledge high above Bear Swamp lower dam show how one mineral (probably calcite) has been dissolved out of the bedrock. Broken pieces of rock can range in scale from the 25 ton chunk dislodged from the ledge at Bear Swamp Power Station in 1984 to tiny particles the size of a sand grain. These processes, and the carrying away of the resulting debris by streams and rivers, are called erosion. Although usually slow, erosion is unrelenting; given twenty or thirty million years, it can wear down and carry away thicknesses of rock measured in miles from areas previously elevated by plate tectonic events. Occasionally erosion is dramatic like the 1938 landslides along the Deerfield. Streams and rivers carry this debris in the form of mud, sand, and silt and eventually deposit it as layers of sediment in the oceans along the margins of the continents. The sand on ocean beaches is the most obvious example of such sediment. The process of depositing this sediment is called sedimentation. Eventually, as these layers of mud and sand are buried beneath subsequent layers, they become compacted and converted to rocks called shale and sandstone. Shales and sandstones caught between two colliding plates can be transformed by heating and squeezing into rocks called schist and quartzite, which characterize much of the town of Rowe today.

Glacier ice is an effective carrier of dirt and debris. As the ice moves slowly over the ground surface, it picks up clay, sand, pebbles, and even huge barn- and house-sized boulders and incorporates them into the ice. When the glacier ice melted, huge quantities of water flooded many of the streams of the area. Terrace deposits laid down by these meltwater streams along the Deerfield River are as much as 300 feet above the present stream level and attest to the volume of this flow. The large and small boulders that we see today perched on the ground in the middle of a field or in the woods were dropped there by the melting glaciers. One of the largest of these boulders is high on the hillside west of Zoar Road above the watering trough. Another is above the switching station off Tunnel Road. In some cases we can tell from the character of the rock that a boulder must have been carried by the ice for miles from its original source. Even more important to the area are the many deposits of sand and gravel that were formed by these glacial meltwater streams which transported, sorted, and deposited dirt freed from the melting ice. These sand and gravel deposits formed as terraces or sand bars, like Rowe's Beaver Island, along streams such as the Deerfield and in temporary ponds formed against residual patches of ice, long since melted. Good examples are the sand and gravel deposits along Cyrus Stage Road and Stone Hill Road in the northeast corner of town. The sand and gravel deposits along the Deerfield in Charlemont continue to be a valuable economic resource.

The spectacular topography along the Deerfield resulted from a combination of events. The course of the river through Rowe in pre-glacial times is believed to have been essentially the same as it is today. As the ice advanced southward across Massachusetts, it scoured out the valley, deepening and modifying it slightly. When the glaciers melted, great quantities of water were sluiced down the Deerfield further steepening and deepening the valley and depositing patches of terrace gravel along the valley walls.

"Erratic" Boulder dropped by the Glacier near the Switching Station in west Rowe.

Pelham, Davis Mine, and Steele Brooks, the other principal streams in the town, are similarly believed to have existed in much their present form in pre-glacial times. Scouring by the ice and flushing by glacial meltwater probably caused a few minor divergences from their former courses, but we would have little trouble recognizing the present drainage pattern were we to see Rowe as it looked 100,000 years ago. This drainage system itself proved to be of great economic value to Rowe providing water power for early mills along the brooks and water capacity for the present day hydro, pumped storage, and atomic electric generating stations on the Deerfield.

With the melting of the ice sheet, Rowe was left basically as it is today with the obvious exceptions of vegetation cover and the effects of man. The talc and pyrite deposits that have played such an important role in the early history of man's time in the town owed their very existence to the happenstance of the level of erosion. Had these particular deposits been originally formed a bit higher in the earth's crust, they would have been eroded away long ago. Had they been formed a bit lower, they might never have been discovered. This particular erosion pattern also left the rugged terrain of a "hill town" which would prove to be both an attraction and a discouragement to its settlement. Looking back at the earliest history of Rowe, it seems certain that the first 600 million years of geologic history have profoundly affected the last two hundred years of Rowe's human history.

GLOSSARY OF TERMS

Amphibolite: A dark-green to black rock composed of variable proportions of hornblende (a mineral typically forming black needles) and white feldspar. Formed by the metamorphism (heat and pressure) of basalt (a dark-colored volcanic rock).

Crust: The outer layer or skin of the earth. Continental crust is about 30 miles thick and is composed largely of light-colored rocks like granite. Ocean crust (beneath the oceans) crust is only about 10 miles thick and is composed largely of dark-colored rocks like basalt (dark volcanic rock).

Gneiss: A variably colored rock generally characterized by alternating layers or bands an inch or two thick of light-colored granular quartz-feldspar rock and dark-colored schist containing lots of biotite (black mica) and/or hornblende (a mineral typically forming black needles that is a common constituent of metamorphosed volcanic rocks). Formed most commonly by the metamorphism (heat and pressure) of some volcanic rocks.

Greenstone: A green to dark-green rock composed predominantly of feldspar, chlorite (green mica) and amphibole (a group of dark-green to black minerals, including hornblende, commonly in needlelike crystals) formed by metamorphism of basalt (dark volcanic rock).

Metamorphism: The process wherein a rock recrystallizes new mineral grains, commonly with a new rock texture, as a result of being subjected to high temperatures and pressures after burial in the earth's crust.

Orogeny: A geologic event in which the rocks of an area are folded, faulted, and commonly metamorphosed (heated and squeezed). The major orogenies affecting the rocks of the Appalachian mountain chain are generally believed to have resulted from two continental (land) masses colliding and crushing the rocks that were on their margins and in between the two continents.

Quartzite: A light-colored (white, pale-gray, tan) fine-grained (commonly the texture of lump sugar) rock composed predominantly of grains of quartz with lesser amounts of feldspar and black (biotite) and white (muscovite) mica. Formed by the progressive metamorphism (heat and pressure) of sand converting to sandstone to quartzite.

Pyrite: A brass-colored, metallic-appearing mineral composed of iron and sulfur. Sometimes called Fool's Gold. Mined in Rowe as a source of sulfur for sulfuric acid. No longer used as an ore of sulfur.

Schist: A rock composed of at least 25 percent of grains of platy minerals, most commonly black (biotite), white (muscovite), or green (chlorite) mica, oriented parallel to each other so that the rock tends to break apart in slabs or sheets. The rock can be white, gray, or green depending upon the proportions of the constituent minerals. Formed by the progressive metamorphism (heat and pressure) of mud or sandy mud converting to shale to mica schist.

Serpentine: A mineral, $Mg_3Si_2O_5(OH)_4$. Name also applied to a rock made predominantly of the mineral. Commonly associated with talc in soapstone where it forms the darker green, harder constituent.

Soapstone: A rock composed predominantly of talc with small amounts of serpentine.

Talc: A mineral, $Mg_2Si_4O_{10}(OH)_2$. Name also applied to a rock made predominantly of the mineral. Used to make talcum powder as well as in the manufacture of paint, ceramics, etc.

Thrust fault: A fault or break in the rocks with a gently inclined surface along which the rocks on one side of the fault have been shoved up and over the rocks on the other side of the fault.

Greenfield Recorder Photo 1984

Man Harnesses Rowe's Drainage System For Power. Sherman Pond on the Deerfield River provides cooling waters for Yankee Atomic Electric Company's nuclear generating station, and water power for New England Power Company's tiny Sherman Hydro Electric Station. Earliest user of the water was the paper mill downriver at Monroe Bridge. Notice natural terrace formations above the River.

About the Author: Norman L. Hatch, Jr. was born in 1932 and has received his BA, MA, and PhD all from Harvard University. He is a Geologist with the U.S. Geological Survey and from 1973-1978 served as Chief, Branch of Eastern Regional Geology. His professional specialty and chief interest is the structure and tectonics of the Northern part of the Appalachian Mountain System. His specific project studies have been in Maine, New Hampshire, Vermont, and Western Massachusetts. Based in Rowe for three summers, Mr. Hatch studied and mapped the rocks of western Massachusetts between 1963 and 1972. He is author or co-author of about fifty published maps and reports, primarily on the rocks of New England.

of a stop-and-go, speed-up and slow-down process, we can estimate that the overall rate of erosion of the rock that once covered Rowe was a few hundredths of an inch per 100 years. A classic example of slow but steady wins the race!

B & M Railroad Photo

Dramatic Erosion. 900 foot Landslide from Negus Mountain into the Deerfield River during the 1938 Hurricane.

The final chapter in the geologic history of Rowe had to wait until, in terms of 600 million years of total history, virtually the last few moments. The Pleistocene epoch, or glacial age, began about 2 million years ago; during this time northern North America was repeatedly covered by huge ice sheets. However, evidence for only the last two such ice sheets is found in western Massachusetts. The first of these two ice sheets advanced southward and covered Rowe about 85,000 years ago. It was thick enough to cover even the highest hills of Rowe with a quarter mile or more of ice, and probably was significantly larger than the great ice sheet that presently covers most of

Greenland. As this ice sheet advanced over New England, it acted like a huge bulldozer scraping loose soil and rock and incorporating them into the ice itself or smearing them out over the landscape as the till or hardpan that forms the dirt or soil over at least 90 percent of Rowe today (all of the town except those areas on the geologic map shown as covered by deposits of sand and gravel).

Adams Mountain and Glacial Depression where collected deposits of sand, gravel, silt and clay. Area known to early settlers as "The Great Swamp" is now man-made Pelham Lake. In Moretown Formation.

A warming trend in the climate melted the ice back sufficiently far north that Rowe reappeared from beneath its icy blanket, but not for long. About 25,000 years ago, the ice had advanced again from the north, and Rowe was completely covered for the second time. Best estimates are that this ice sheet remained over western Massachusetts for about 11,000 years before another warming cycle drove the leading edge of the sheet back to the north uncovering Rowe for the last time about 14,000 years ago, leaving glacial grooves in the bedrock such as the readily observed example north of the power line on the east side of Potter Road.

Rowe Postcard c. 1900

Ledges of Hoosac Formation near Pulpit Rock. The folded near-vertical layering in the rock originally formed as horizontal strata on an ocean floor.

The following paragraphs outline the plate tectonic and sedimentation events and other geologic processes that make up the earliest history of Rowe.

Approximately 600 million years ago, an ancient continent, of which the rocks of the Berkshire Hills a few miles to the west of Rowe are a part, was pulled apart by powerful forces in the earth's crust along a line that roughly traces through the villages of Adams, Hinsdale, Becket, and Otis. An ocean formed between the pulled apart pieces. The area of the present town of Rowe lay beneath that ocean and was the site of deposition of mud and sand (which later hardened to rocks now known as the Hoosac Formation) that eroded off the piece of the pulled-apart continent to the west—what we now call the Berkshire Hills. Rowe and its environs remained beneath the ocean and continued to receive more sediment (now known as the Rowe Schist), at least intermittently, for the next 150 million years.

By about 450 million years ago, the circumstances in the vicinity of Rowe had changed dramatically. A chain of volcanic islands, perhaps analogous to the Aleutians of present times, was advancing slowly westward toward North America, gradually closing up the ocean basin. As the two land masses approached each other, sand and mud were eroded from both the volcanic islands and the Berkshire Hills, and were deposited in the intervening ocean to form the future Moretown Formation. Volcanic ash and lava flows and some black mud derived from the volcanic islands were then deposited in the closing ocean to form the future Hawley Formation. These materials formed thick layers on the black heavy volcanic rock that characterizes the earth's crust beneath the oceans. Probably at the same time, sulfur from the volcanoes combined with iron, copper, and small amounts of other metals in the rocks to form the deposits of the future Davis and Mary Louise mines and related prospects. All of these rocks were then caught in a giant closing vise, the jaws of which were the chain of volcanoes on the east and the ancient continent (North America) represented by the Berkshires on the west.

As the jaws of the visc continued to close, the intervening slab of ocean-floor volcanic rock and its mantle of sediment and volcanic rocks became highly compressed, with spectacular results. First, the combination of great pressure and high temperature (at least a few hundred degrees Fahrenheit), called metamorphism, converted the sand and mud into hard rocks, called quartzite and mica schist, and the volcanic ash and lava flows into gneiss, feldspar schist, and amphibolite. Also at this time, as a result of the compression and the availability of water, most of the ocean-floor rock was converted to serpentine, one of the principal ingredients of soapstone. Second, the compression formed a set of roughly parallel faults which sliced the volcanic rocks of the ocean floor and the overlying metamorphosed sedimentary and volcanic rocks into slabs. The slabs were shuffled westward almost like cards in a deck resting on a surface that sloped very gently to the east. This shuffling not only mixed up the metamorphosed sedimentary and volcanic rocks but also injected slices of the serpentinized ocean-floor rock up into them. Then, late in the period of closing of the the vise, the whole complex of the metamorphosed sedimentary and volcanic rocks and the injected slices of serpentine rocks was shoved westward up onto the eastern margin of the North American continent along a giant fault called the Whitcomb Summit thrust fault. The jaws of the vise had completely closed, the ocean had disappeared, and the two land masses, the North American continent and the volcanic island chain, had become welded together. This series of truly cataclysmic events, which probably extended over a period of 10 to 20 million years, is known as the Taconic orogeny. At the end of the Taconic orogeny, around 440 million years ago, the rocks that are presently exposed at the earth's surface in Rowe were probably buried a few miles below the surface.

The history of Rowe seems to have been less dramatic for the next 30 million years or so. A long, narrow north-south sea formed off to the east. This sea may or may not have flooded as far west as Rowe; the gray mud and sand (now gray mica schist) that were deposited in it have not been recog-

GEOLOGIC MAP OF ROWE

Map is simplified from geologic maps of the Rowe and Heath quadrangles published by the U.S. Geological Survey as U.S. Geological Survey Quadrangle Maps #642 and 735. The rocks are divided into four bedrock geologic formations. These are, from oldest to youngest, the Hoosac, Rowe, Moretown, and Hawley. Although ledge, or bedrock, is exposed in only a small percentage of the town, the available information enables a reasonable prediction of the distribution of the different formations beneath the surficial cover of sand, soil, etc. The four formations shown on the map contain some rock types in common, and in many places, grade into one another. Thus, the distinction between the formations may not be obvious, even after looking at a number of bedrock exposures. Covering the bedrock throughout all but one or two percent of the town is a variable thickness of unconsolidated material. Where this unconsolidated material is sand or gravel deposited by streams, it is distinguished on the map by a special pattern. Where this pattern is not shown, it should be assumed that some thickness of till or hardpan, deposited by glaciers, overlies the designated bedrock formation. For additional detail, see the Geologic Quadrangle Maps cited above.

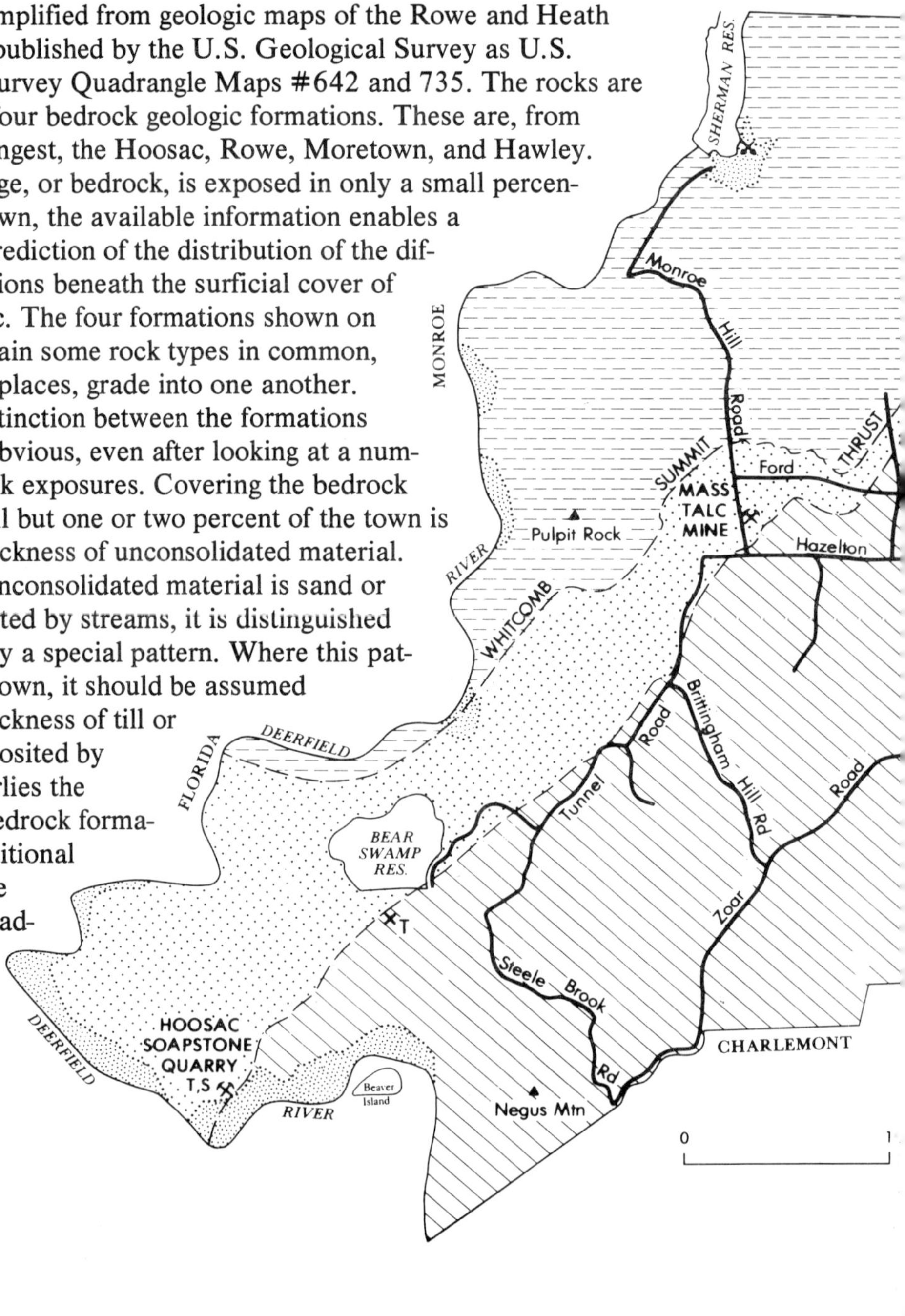